The LIBRARY of LANDFORMS™

MOUNTAINS

Isaac Nadeau

The Rosen Publishing Group's

PowerKids Press™

New York

To the Cuzzies

Published in 2006 by The Rosen Publishing Group, Inc.
29 East 21st Street, New York, NY 10010

First Edition

Editor: Rachel O'Connor
Book Design: Elana Davidian

Photo Credits: Cover, title page © Paul Chesley/National Geographic/Getty Images; p. 4 © Richard I'Anson/Lonely Planet Images; p. 8 (top) © Nick Tapp/Lonely Planet Images; p. 8 (bottom) © Yann-Arthus Bertrand/Corbis; p. 11 © Jim Wark/Lonely Planet Images; p. 12 © Greg Vaughn/Getty Images; p. 15 (top) © Thomas Downs/Lonely Planet Images; p. 15 (bottom) © Corey Rich/Lonely Planet Images; p. 16 (top) © Randy Olson/National Geographic/Getty Images; p. 16 (bottom) © Gary Braasch/Corbis; p. 19 © David Hiser/Getty Images; p. 20 (top left, bottom) © Sumio Harada/Minden Pictures; p. 20 (top right) © John Condrad/Corbis.

Library of Congress Cataloging-in-Publication Data

Nadeau, Isaac.
  Mountains / Isaac Nadeau.
      p. cm. — (Library of landforms)
  Includes index.
  ISBN 1-4042-3127-7 (lib. bdg.)
  1. Mountains—Juvenile literature. I. Title.

  GB512.N34 2006
  551.43'2—dc22
                                    2005001558

Manufactured in the United States of America

CONTENTS

The tallest mountain on Earth is Mount Everest. It measures 29,035 feet (8,850 m) tall. Mount Everest is part of the Himalayan range in southern Asia. *Himalaya* is a Sanskrit word meaning "home of snow." The mountain peaks of the Himalayan range are covered in snow year-round.

What Is a Mountain?

Mountains are Earth's highest **landforms**. They are like hills but are taller and larger. Some scientists define a mountain as being more than 2,000 feet (610 m) high. Mountains usually have steep sides and a high point, called a peak or summit. Most mountains come in groups, called mountain ranges. For example, the Cascade Range includes many mountains that cover about 700 miles (1,127 km) from northern California to southwestern Canada. There are mountains on every **continent** and in every ocean. Mountains can be islands, such as the mountains that make up the Hawaiian Islands. Mountains can be found in the middle of continents, such as the Rockies in North America. Some mountains, such as the Appalachian Mountains in eastern North America, are hundreds of millions of years old. These mountains have been **eroded** over time. Much of their rock has been washed away by rain, melting snow, and streams. They are not as high as they once were. Other mountains, such as Mauna Loa in Hawaii, are young and still growing. Mauna Loa is about one million years old, which is considered young for a mountain!

BUILDING MOUNTAINS

To understand how mountains are formed, it is important to know about Earth and its moving parts. Earth is a giant sphere, or ball. The continents and the ocean floor form a rocky crust around Earth. Earth's crust is divided into many large pieces, called plates. These plates float on top of a deeper layer of Earth, called the mantle. It is made up of rock so hot that it is soft. The hottest rock in the mantle is **magma**, which rises toward the crust. As the hot, liquid rock from the mantle presses against the bottom of the crust, it causes the plates to move. In some places two plates move away from one another. In other places the plates push against each other. Most mountains on Earth are formed where two plates are pushing against each other. Even though the plates move very slowly, there is a lot of force when they meet. In some cases the plates **collide**, and the land is forced upward. This action forms mountains. The Himalayas in Asia and the Alps in Europe were formed in this way. For billions of years, the movements of Earth's plates have built new mountains and the forces of erosion have worn them away.

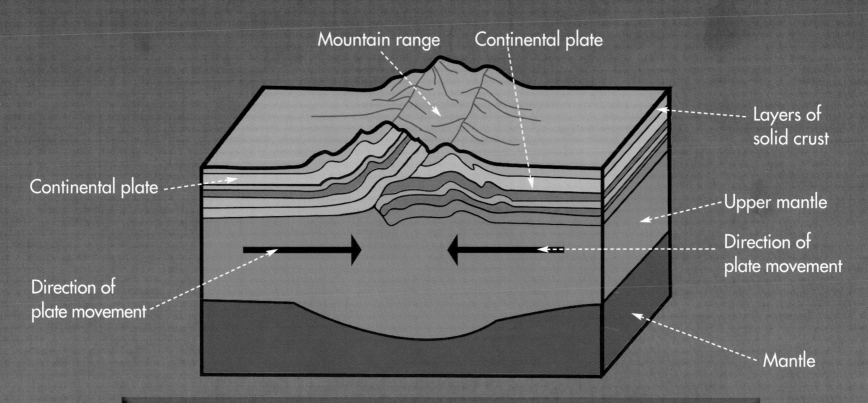

Mountain range

Continental plate

Layers of solid crust

Upper mantle

Direction of plate movement

Continental plate

Direction of plate movement

Mantle

In this diagram you can see how mountains are formed when the plates of Earth's crust push up against each other.

4.5 BILLION YEARS AGO:
Earth is formed.

500–225 MILLION YEARS AGO:
Continents move together to form one giant continent, known as Pangaea. The Appalachian Mountains are formed, reaching heights of 20,000 feet (6,096 m) or more.

165 MILLION YEARS AGO:
The Atlantic Ocean begins to form, as Pangaea breaks up. The continent of Africa separates from South America. North America and Europe separate.

45 MILLION YEARS AGO:
Two plates collide, beginning the formation of the Himalayan mountains in southern Asia.

1 MILLION YEARS AGO:
The Big Island in Hawaii begins to form, as lava shoots up from the ocean floor.

TODAY:
The Himalayas continue to grow at a rate of about 1.2 inches (3 cm) per year as the plates press together.

Top: Monte Sagro in Tuscany, Italy, is an example of a mountain that has the metamorphic rock marble. *Right:* Here you can see a mountain in northwestern Australia that is being mined for diamonds.

Folded Mountains

The most common type of mountain is a folded mountain. When two plates push against each other, the rock on each plate is forced upward, forming mountains. The levels of rock in the mountains appear to be folded. This is how this type of mountain gets its name. The tallest mountains on Earth, the Himalayas, were formed in this way. The Himalayas form the northern border of India. India is on a plate that crashed into the southern coast of Asia beginning about 45 million years ago. The plate's crust was forced upward, creating the high mountains found in India, Nepal, and Tibet today. Today the Indian plate is still moving northward. It is moving very slowly but with a lot of force. As a result the Himalayan mountains continue to grow taller.

An interesting thing happens to rocks when folded mountains are being formed. When two plates collide, the rocks in the plates are put under great heat and **pressure**. This causes the **igneous rocks** or **sedimentary rocks** in the plates to change. The rocks that they change into are called metamorphic rocks. Marble is an example of a metamorphic rock.

Fault-block Mountains

Another type of mountain is a fault-block mountain. A fault-block mountain is formed when plates collide along fault lines. Faults are cracks in Earth's crust. Faults can form when colliding plates stop moving or change direction. This results in the crust stretching, or spreading out, like an accordion being opened up. As the crust stretches, many faults form. Sometimes huge blocks of Earth's crust are pushed upward along these faults, forming mountains. Other blocks drop down through the faults, forming valleys.

The Basin and Range Province in the southwestern United States has many fault-block mountains and valleys. They were formed millions of years ago when two plates were pushed together along the western edge of North America, forming a high **plateau**. Then about 30 million years ago, the western part of North America began to move toward the northwest. This caused the land to stretch out and form cracks, along which the mountains were formed. On a **relief map,** the Basin and Range Province appears as many long mountain ranges running north and south, separated by valleys.

The Guadalupe Mountains in Texas are fault-block mountains. Shown here is El Capitan, one of the peaks found at the southern end of this beautiful mountain range. It measures 8,085 feet (2,464 m). The highest point in Texas is found in this mountain range. It is Guadalupe Peak, which stands at 8,749 feet (2,667 m) high.

Mauna Loa is a volcanic mountain in Hawaii. The islands of Hawaii are formed at a hot spot. Hot spots are weak places found in Earth's crust. They are often found in the middle of a plate. At a hot spot, magma rises through the weak place in the crust. The magma cools at the surface and hardens to form rock. This rock builds higher and higher to form mountains.

VOLCANIC MOUNTAINS

Volcanic mountains are another type of mountain found on Earth. Below Earth's crust is the mantle, which is made up of magma. When magma pushes up through the crust, it is called lava. When lava flows onto Earth's surface, it cools and hardens to form rock. As more and more lava flows to the surface and hardens, the rock builds up and mountains are formed. Volcanic mountains are usually formed at the place where two plates meet. When one plate is heavier than the other, it pushes the lighter plate beneath the surface, into the mantle below. As the plate pushes into the mantle, its rock heats up and becomes magma. The magma rises toward the surface, pushing up the land and forming volcanoes. The Andes Mountains on the west coast of South America are an example of this type of volcanic mountain. Many volcanic mountains are found around the edge of the Pacific Ocean. The edge of the Pacific Ocean is often called the Ring of Fire because of all the volcanoes around it.

Mauna Kea is a volcanic mountain in Hawaii. Rising from the sea, it measures 13,796 feet (4,205 m) above sea level. However, if you were to measure Mauna Kea from its base at the bottom of the ocean, it would stand at 33,465 feet (10,200 m) tall. That is 4,430 feet (1,350 m) taller than Mount Everest, which is known as Earth's tallest mountain!

Dome Mountains

A dome mountain gets its name because it is shaped like a great big dome. Dome mountains are usually not as high as folded mountains. The forces that create them are not strong enough to lift them up as high. Dome mountains are formed when hot magma from Earth's mantle pushes up beneath the surface but does not break through to the surface. As it rises the magma pushes the crust upward as though forming a volcano. However, not enough magma is present to break through the surface, and the rock cools and hardens underground. The result is a dome-shaped mountain. Half Dome, a mountain in the Sierra Nevada range in California, is an example of a dome mountain. Half Dome is made of granite, which is a type of light-colored rock that was once molten but cooled underground. Granite is a very hard rock and is slow to erode. When Half Dome was first formed, there were many layers of softer sedimentary rock above it. Over the years this softer rock was washed away, leaving the hard granite dome behind.

Here you can see Half Dome in Yosemite National Park in California. *Inset:* The Black Hills in South Dakota are an example of dome mountains. Although they are called the Black Hills, they are really mountains. They reach heights of more than 7,000 feet (2,134 m).

Top: The Appalachian Mountains, shown here, are much rounder and softer in appearance than the sharp peaks of younger mountains, such as the Himalayas. *Bottom:* Mountains are changing all the time. For example, When Mount St. Helens in the Cascade Range in Washington erupts, or explodes, it erupts so forcefully that entire sides of the mountain are blasted to bits.

Mountains over Time

Without the formation of new mountains, Earth's surface would finally be worn flat. This is because of erosion. Mountains are eroded when wind, water, ice, rain, and **gravity** work together to wear away rock and drag it downhill. As drops of rain gather into streams flowing down the mountain, the water cuts against the rock, carrying away tiny bits of rock. These bits of rock help scrape even more rock away. When the Appalachian Mountains were formed, they reached heights of more than 20,000 feet (6,096 m). As a result of erosion during the last 160 million years, these mountains have been worn down to less than 7,000 feet (2,134 m). As Earth's plates continue to move and collide, mountains will continue to form and grow, even as others are worn away. Volcanic mountains that continue to erupt, such as the Aleutian Range in southeastern Alaska, add new rock to their mountains every year. All over the world, mountains are changing, growing, and **shrinking** every day.

THE ROCKY MOUNTAINS

The Rocky Mountain range includes folded mountains, fault-block mountains, dome mountains, and volcanic mountains. In the millions of years since the Rocky Mountains were formed, they have been eroded by rain, snow, ice, and streams.

The Rocky Mountains are the most famous mountains in North America. The Rocky Mountains are a large mountain range that measures about 3,000 miles (4,828 km) from northwestern Canada to New Mexico. They were formed between 60 and 35 million years ago. Many **geologists** believe that a large plate to the west was pushed beneath the North American continent. This plate slid underneath the continent, and volcanic mountains were formed near the coast. In time the plate sliding underneath the continent moved hundreds of miles (km) inland, until the entire plate slid underneath the continent. At this point the bottom plate pushed up against the plate above it. This caused a huge part of the continent to be lifted upward, forming a giant plateau called the Colorado Plateau. To the east of the Colorado Plateau, the rise in the land caused the surface to break and lift, creating Colorado's Rocky Mountains.

The state of Colorado is known as the Rocky Mountain State. There are 54 peaks that are more than 14,000 feet (4,267 m) high in Colorado. Here you can see a group on horseback riding through Colorado's Rocky Mountains.

Top Left: Blue clematis flowers are found in the Canadian Rockies. They are usually found in the montane and subalpine zones. *Top Right:* Mountain goats can be found in the alpine zone. *Bottom:* A few birds, such as the bald eagle, shown here, can be seen above the treeline of the Rocky Mountains.

The Rocky Mountains are home to thousands of kinds of plants and animals. Plant and animal life on a mountain changes as you move from the bottom toward the top. This is mainly because **temperatures** get colder the higher up you go. Life in the Rocky Mountains can be divided into different life **zones**. Near the bottom of the Rocky Mountains, the plants include cottonwoods growing along streams and juniper trees in drier areas. Animals include jackrabbits and deer. The next zone is called the montane life zone. It is found at heights of between 6,000 and 9,000 feet (1,829–2,743 m). The major trees here include douglas fir and ponderosa pine. Mountain lions and red squirrels are among the animals found here. The subalpine zone is between about 9,000 and 11,000 feet (2,743–3,353 m). Here spruce and fir are the main trees. Animals include snowshoe hares and lynx. The highest point on a mountain where trees can grow is called the treeline. The alpine zone is above the treeline. There are still many plants and animals that live above the treeline. Most alpine plants, such as alpine daisy, grow low to the ground, where it is warmest.

MOUNTAINS AND PEOPLE

In the mountains of China, people grow rice using a system of farming called terracing. Terraces are fields cut into the mountainsides like staircases. This allows the farmers to grow their crops on flat ground, to prevent soil and water from being washed downhill, and to direct water easily from one terrace to the next.

About one-fifth of Earth's surface is made up of mountains. One out of ten people in the world lives in the mountains. However, the steep slopes and cold **climates** can make mountains hard places in which to live. The people who live on mountains have found special ways to **adapt**. For example, the mountains of Nepal in the Himalayas are very steep. The Sherpa people of Nepal have **developed** a system of mountain paths. This helps them travel from village to village and from the highlands to the lowlands. They use the yak for its milk and meat. A yak is a type of furry mountain cow. The yak also carries firewood, salt, and other goods from place to place.

Mountains are popular places for people to visit. Rock climbing, downhill skiing, biking, and hiking are a few of the reasons people visit mountains. As places to live, study, and explore, mountains have been important to people for thousands of years. They will continue to be special places for thousands of years to come.

Glossary

adapt (uh-DAPT) To change to fit new conditions.

climates (KLY-mits) The kinds of weather a certain area has.

collide (kuh-LYD) To crash together.

continent (KON-teh-nent) One of Earth's seven large landmasses.

developed (dih-VEH-lupt) Worked out.

eroded (ih-ROHD-ed) Worn away slowly.

geologists (jee-AH-luh-jists) Scientists who study the form of Earth.

gravity (GRA-vih-tee) The natural force that causes objects to move toward the center of Earth.

igneous rocks (IG-nee-us ROKS) Hot, liquid, underground minerals that have cooled and hardened.

landforms (LAND-formz) Features on Earth's surface, such as a hill or a valley.

magma (MAG-muh) A hot, liquid rock underneath Earth's surface.

plateau (pla-TOH) A broad, flat, high piece of land.

pressure (PREH-shur) A force that pushes on something.

relief map (rih-LEEF MAP) A map that shows how high or low places are.

sedimentary rocks (seh-deh-MEN-teh-ree ROKS) Layers of gravel, sand, silt, or mud that have been pressed together to form rocks.

shrinking (SHRINK-ing) Becoming smaller due to heat, cold, or wetness.

temperatures (TEM-pruh-cherz) How hot or cold things are.

volcanic (vol-KA-nik) Having to do with a volcano.

zones (ZOHNZ) Areas that have special conditions or uses.

INDEX

WEB SITES